S0-BCO-131

DISCARDED

DISCARDED

LET'S·READ·AND·FIND·OUT SCIENCE®

STAGE 2

Down Comes The

RAIN

by Franklyn M. Branley
illustrated by James Graham Hale

HarperCollinsPublishers

Special thanks to Don W. Hen for his expert advice.

The illustrations in this book were done in pen-and-ink with watercolor washes.

The *Let's-Read-and-Find-Out Science* book series was originated by Dr. Franklyn M. Branley, Astronomer Emeritus and former Chairman of the American Museum–Hayden Planetarium, and was formerly co-edited by him and Dr. Roma Gans, Professor Emeritus of Childhood Education, Teachers College, Columbia University. Text and illustrations for each of the books in the series are checked for accuracy by an expert in the relevant field. For more information about Let's-Read-and-Find-Out Science books, write to HarperCollins Children's Books, 10 East 53rd Street, New York, NY 10022.

HarperCollins®, ✥®, and Let's Read-and-Find-Out Science® are trademarks of HarperCollins Publishers Inc.

Library of Congress Cataloging-in-Publication Data
Branley, Franklyn Mansfield, date
   Down comes the rain / by Franklyn M. Branley ; illustrated by James Graham Hale.
      p.   cm. — (Let's-read-and-find-out science. Stage 2)
   ISBN 0-06-025338-X.(lib. bdg.). — ISBN 0-06-025334-7 —ISBN 0-06-445166-6 (pbk.)
   1. Rain and rainfall—Juvenile literature.   2. Clouds—Juvenile literature.   [1. Rain and rainfall.
2. Clouds.]   I. Hale, James Graham, ill.   II. Title.   III. Series.
QC924.7.B694   1997   96-3519
551.57—dc20   CIP
                                                                                                              AC

Typography by Elynn Cohen
1   2   3   4   5   6   7   8   9   10
❖
Newly Illustrated Edition, 1997

T 2108332

Down Comes The Rain

Rain comes from clouds.

It comes from big clouds

and little clouds.

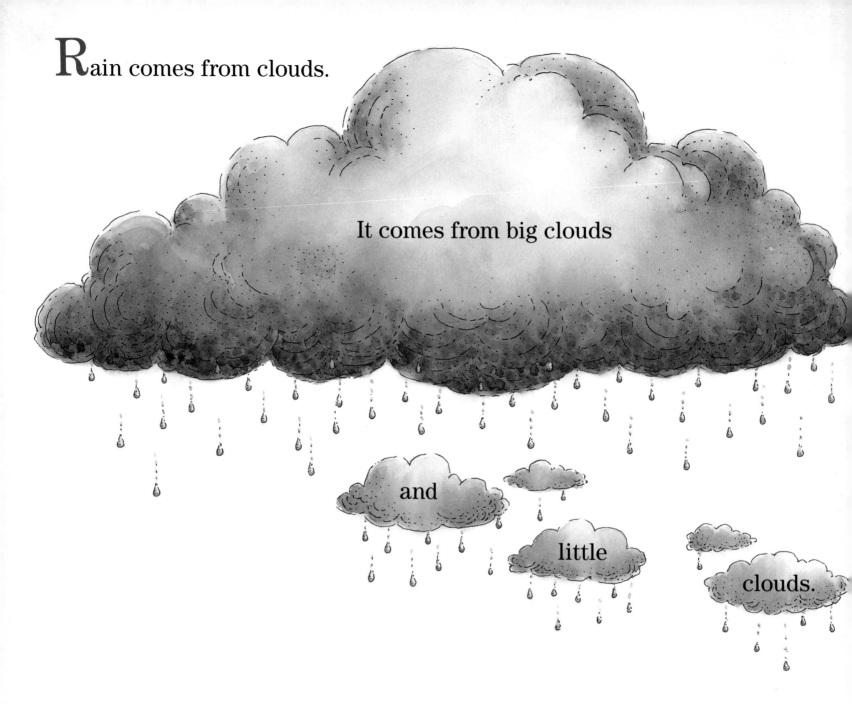

4

It comes from black clouds, white clouds,

and gray clouds.

All clouds—big ones and little ones, gray ones and white ones—are made of billions of tiny drops of water. The drops are called droplets, because they are so small.

If this is the size of a drop of water,

a droplet would be just a tiny speck, even smaller than this one.

Water droplets come from water vapor. Water vapor is a gas.

There's always water vapor in the air, but you can't SEE it...

...can't SMELL it...

...and you can't FEEL it.

7

Water vapor is made when water evaporates. That means the water changes from a liquid to a gas.

In the morning, put a teaspoon of water in a saucer...

By that night, it may have EVAPORATED into the air!

hen wet clothes
g on the clothesline,
ater in them evaporates.
The heat from the sun
hanges the water drops
and droplets into
water vapor.

Just like the heat
from the stove changes
water in the kettle to water vapor.
If you heat it long enough,
all the water boils away.
The water vapor
goes into the air.

9

Most of the water vapor in the air comes from lakes, rivers, and oceans. It comes from the leaves of plants, and from the wet ground.

Heat from the sun causes the water to evaporate.

The water changes from a liquid to a gas...

...and the water vapor goes into the air.

When you breathe out, you put water vapor into the air. Usually, you cannot see the water vapor. But it is there.

Sometimes, if it's COLD, you can see your breath.

That's because the water vapor condenses.

It changes from a gas to a little cloud.

When cows, horses, dogs, and cats breathe out, they put water vapor into the air, too. On a cold day, the water vapor changes to droplets and makes little clouds that you can see.

13

You can make water vapor change to water.

Put a lot of ice into a glass of water.

As the glass gets colder, the outside of the glass gets WET.

14

Water vapor in the air is condensing on the glass.

There may be so much condensation that the glass drips.

Sometimes the glass stays dry. That means there is not much water vapor in the air.

The air holds the water vapor. Breezes carry it from one place to another.

Much of the vapor moves up and away from the earth.

Air above the earth is <u>always</u> cold. The higher you go, the COLDER it gets.

When air gets cold enough, the water vapor in it condenses.

The vapor changes to water droplets. The water droplets make clouds.

17

When clouds are THIN and Wispy, they are holding only a Little water.

When clouds are THICK and DARK they are holding much more water.

A single droplet is so small you cannot see it. But you can see a cloud. That's because there are millions and millions and millions of water droplets in a cloud.

Inside the clouds, droplets join together to make drops.

When clouds can no longer hold them, the drops fall to the earth.

The sky is full of them. They fall through the air and splatter on the ground.

They are raindrops.

Sometimes there are only a few small raindrops
that fall slowly. It is drizzling.

Sometimes there are lots of big drops that fall very fast. Now it is pouring.

Sometimes the drops in clouds freeze. These raindrops become ice drops. This can happen even on a hot summer day.

Some clouds may be higher than most airplanes ever go.

The higher the clouds, the colder they are.

That's because the clouds and water droplets are high above the earth.
Many clouds are so high that it is freezing cold.

In these high, cold clouds, water vapor changes to droplets, and the droplets change to drops. The drops freeze into ice.

Inside the cloud, these tiny bits of ice start to fall...

FREEZING COLD!

water vapor → droplets → water drops → ICE drops

air currents

Earth

...but they don't always fall out of the cloud. Instead, they may be carried upward by air that is moving away from the earth.

As they are carried upward, more water collects on the tiny bits of ice. When that water freezes, the drops of ice have another layer on them.

air currents

The ice drops are now heavier, so once more they fall toward the earth. But air moving away from the earth may carry the ice drops upward again. Higher and higher they go, and another layer of ice freezes onto them.

EA

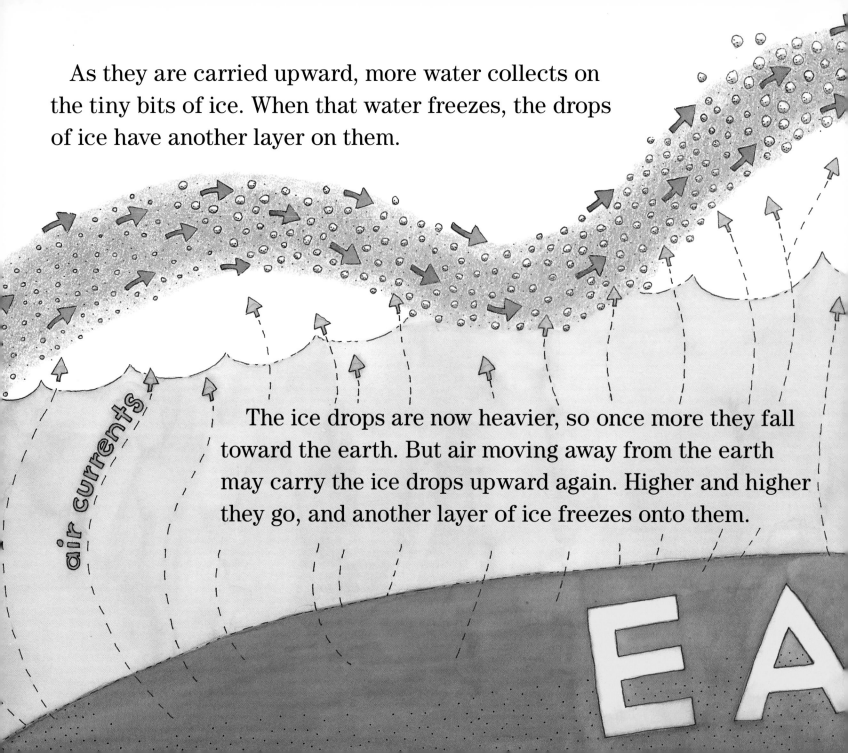

The ice drops get heavier and heavier. They get so heavy that the air can no longer carry them upward.

So the ice drops fall to the earth.

It is raining ice.

Yikes!

25

The ice drops are called hailstones. They may be the size of your fingernail...

... or they may be as big as golf balls, or even BIGGER.

In 1970, hailstones as big as softballs fell on Kansas.

Hailstorm Flattens Crops in Kansas

Fields of corn were FLATTENED by the hailstones!

Hailstones are not stones. They're pieces of ICE. So when it hails, go inside so you're not hit on the head!

When it stops hailing, go outside and pick up a hailstone.

Break it in two, and you will see the layers of ice:

Water in the clouds makes HAIL.

Water in the clouds makes RAIN.

When it stops raining or hailing, the SUN comes out.

Once more, water EVAPORATE

28

It evaporates from lakes, rivers, and oceans.

It evaporates from the leaves of plants and from the wet ground.

It evaporates from cows and horses, from cats and dogs,

...and from you and me.

The water changes to water vapor.

It's carried up and away from the earth, where the air is cool, or even FREEZING.

When the water vapor cools, it condenses.
The water vapor changes to water droplets,
and all together the droplets make clouds.

Water droplets join together to make water
drops. The drops fall to the earth from the clouds.

31

Once more it is raining.